ENERGY FROM
COAL

by J. P. Press

Consultant: Beth Gambro
Reading Specialist, Yorkville, Illinois

BEARPORT
PUBLISHING

Minneapolis, Minnesota

Teaching Tips

Before Reading

- Look at the cover of the book. Discuss the picture and the title.

- Ask readers to brainstorm a list of what they already know about coal. What can they expect to see in this book?

- Go on a picture walk, looking through the pictures to discuss vocabulary and make predictions about the text.

During Reading

- Read for purpose. Encourage readers to think about coal and energy and the roles they play in our daily lives as they are reading.

- Ask readers to look for the details of the book. What are they learning about coal?

- If readers encounter an unknown word, ask them to look at the sounds in the word. Then, ask them to look at the rest of the page. Are there any clues to help them understand?

After Reading

- Encourage readers to pick a buddy and reread the book together.

- Ask readers to name one reason to use coal and one reason to not use coal. Go back and find the pages that tell about these things.

- Ask readers to write or draw something they learned about energy from coal.

Credits:

Cover and title page, © Satephoto/iStock; 3, © bee_photobee/iStock; 5, © Choreograph/iStock; 6–7, © Artur_Nyk/Shutterstock; 8–9, © DE AGOSTINI PICTURE LIBRARY/Getty; 10–11, © 1968/Shutterstock; 13, © ITAR-TASS News Agency/Alamy; 14–15, © stocknroll, © kyoshino/iStock; 16, © Parilov/Shutterstock; 19, © Schroptschop/iStock; 20–21, © James Marvin Phelps/Shutterstock; 22, © Natalia Darmoroz/iStock; 23BL, © Iuliia Korniievych/iStock; 23BR, © Africa Studio/Shutterstock; 23TL, © Hywit Dimyadi/iStock; 23TM, © ShutterProductions/Shutterstock; 23TR, © Hello_ji/Shutterstock

Library of Congress Cataloging-in-Publication Data

Names: Press, J. P., 1993- author.
Title: Energy from coal / by J.P. Press ; consultant Beth Gambro, Reading
Specialist, Yorkville, Illinois.
Description: Bearcub books edition. | Minneapolis, Minnesota : Bearport
Publishing Company, [2022] | Series: Power up with energy! | Includes
bibliographical references and index.
Identifiers: LCCN 2020051858 (print) | LCCN 2020051859 (ebook) | ISBN
9781647478643 (library binding) | ISBN 9781647478711 (paperback) | ISBN
9781647478780 (ebook)
Subjects: LCSH: Coal--Juvenile literature. | Power resources--Juvenile
literature.
Classification: LCC TP325 .P875 2022 (print) | LCC TP325 (ebook) | DDC
662.6/2--dc23
LC record available at https://lccn.loc.gov/2020051858
LC ebook record available at https://lccn.loc.gov/2020051859

For more information, write to Bearport Publishing, 5357 Penn Avenue South, Minneapolis, MN 55419. Printed in the United States of America.

Contents

Keeping Warm

Brr!

It is cold outside.

But it is warm in our homes.

How can that be?

The answer might be in coal.

We need **energy** to keep our homes warm.

Energy gives things power.

It can make a **heater** work.

We can get energy from coal.

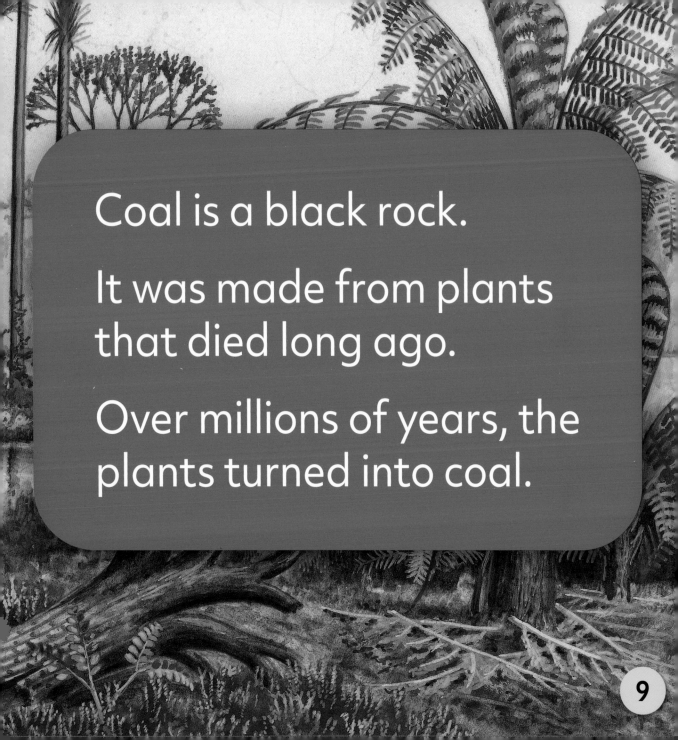

Coal is a black rock.

It was made from plants that died long ago.

Over millions of years, the plants turned into coal.

Most coal is under
the ground.

We need to dig to
find it.

Digging for coal is
called **mining**.

We burn coal to make **steam.**

Then, the steam spins **blades**.

This makes a kind of energy.

13

Energy from coal gives us power for many things.

It can make heat for cooking.

It can also give us light.

We use coal because there is a lot of it.

We know where to find more, too.

It does not take much money to get energy from coal.

But there are bad things about coal energy, too.

We cannot make more coal when we run out.

It can make our air and water dirty.

People are making coal energy cleaner.

For now, we still use coal.

But we will use less as we find more ways to get energy.

Energy from Coal

Follow along to see how coal is made.

1 Big plants died a long time ago.

2 Dirt covered the dead plants.

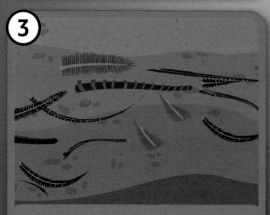

3 Many millions of years went by.

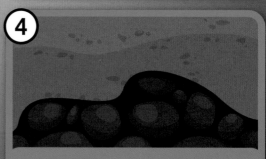

4 Heat and the pushing down from above turned the plants into coal.

Glossary

blades flat spinning parts that are used on some machines

energy power that makes things work

heater a thing used for making warmth

mining digging up things that are underground

steam hot gas

Index

Read More

O'Brien, Cynthia. *Energy Everywhere (Full STEAM Ahead!)*. New York: Crabtree, 2020.

Olson, Elsie. *Coal Energy (Earth's Energy Resources)*. Minneapolis: Abdo, 2019.

Learn More Online

1. Go to **www.factsurfer.com**
2. Enter "**Coal Energy**" into the search box.
3. Click on the cover of this book to see a list of websites.

About the Author

J. P. Press likes to run and read. She thinks we all need to do our part to use safe, clean energy.